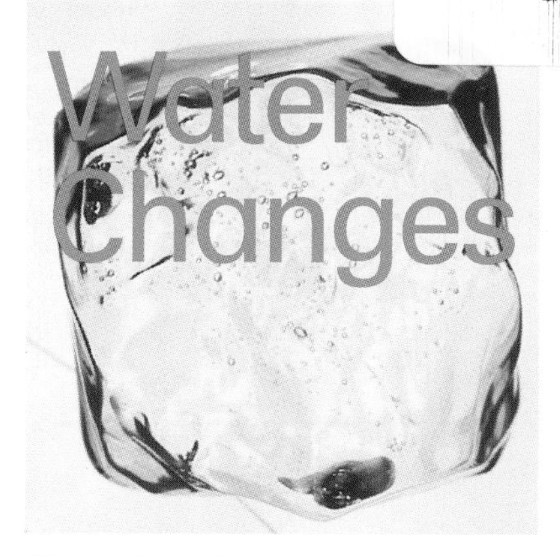

By Josh Weinstein

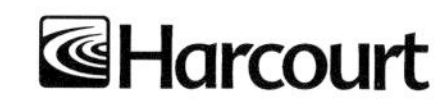

Orlando Boston Dallas Chicago San Diego

www.harcourtschool.com

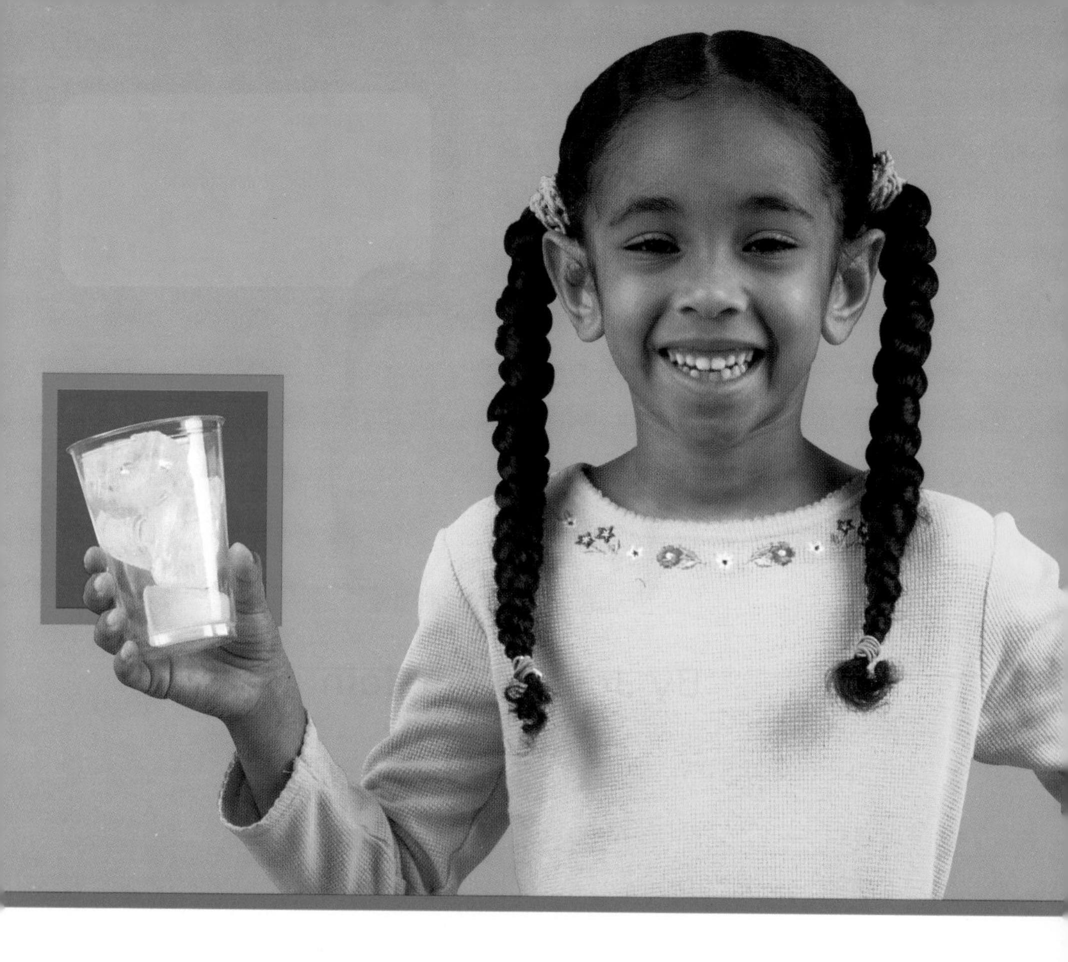

I have ice.

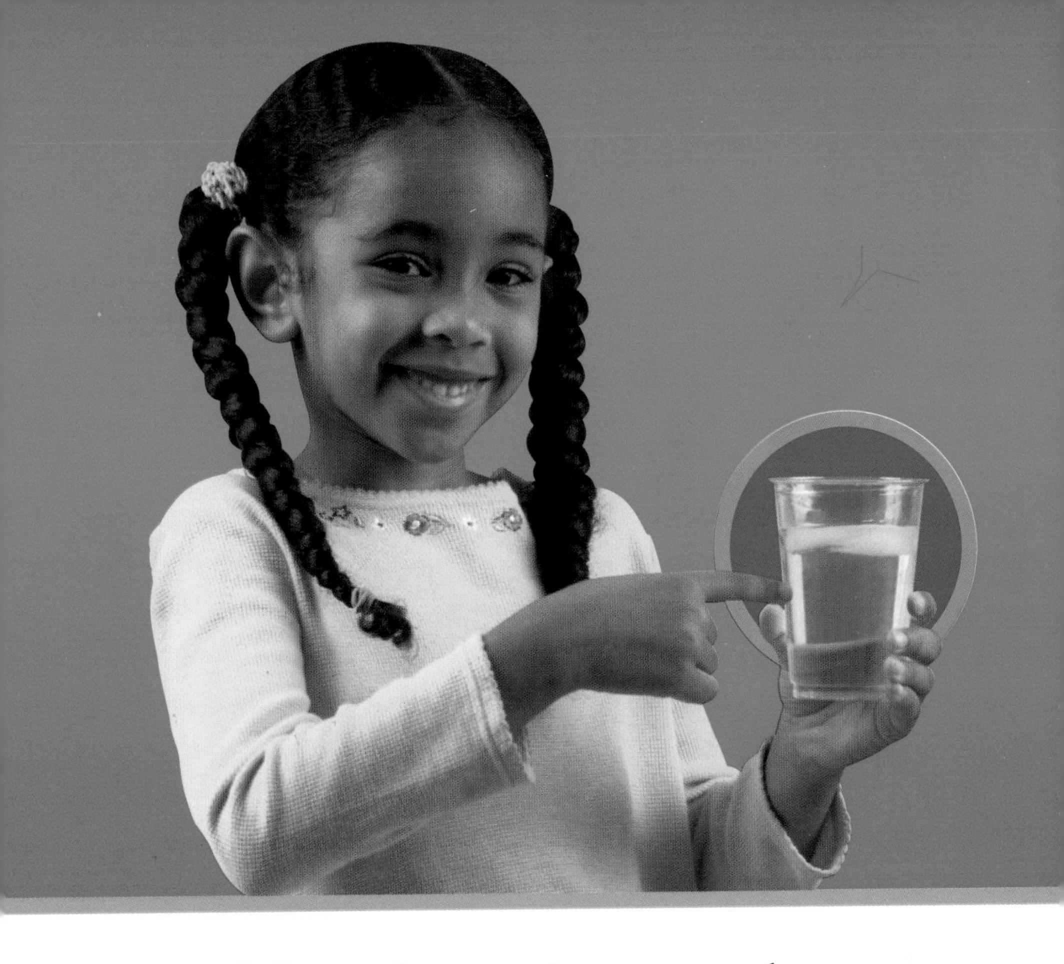

My ice is water.

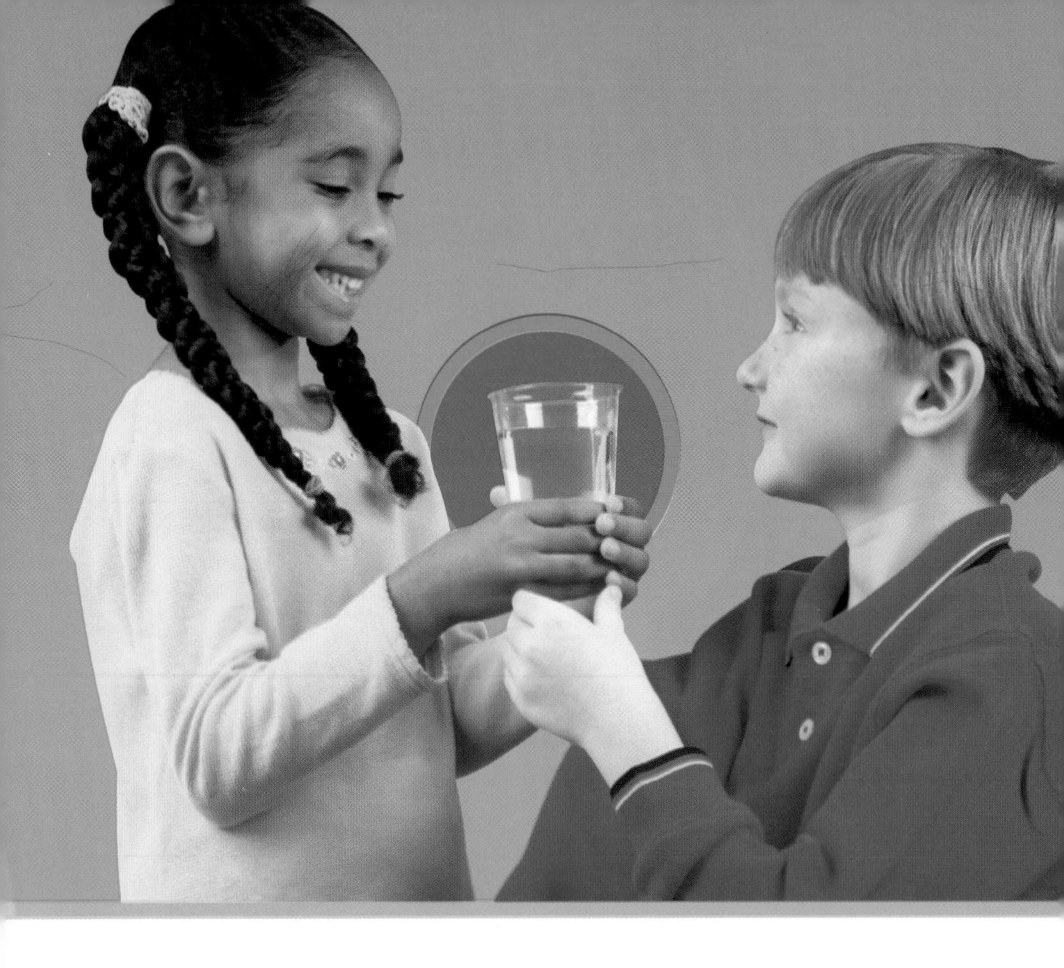

You can have my water.

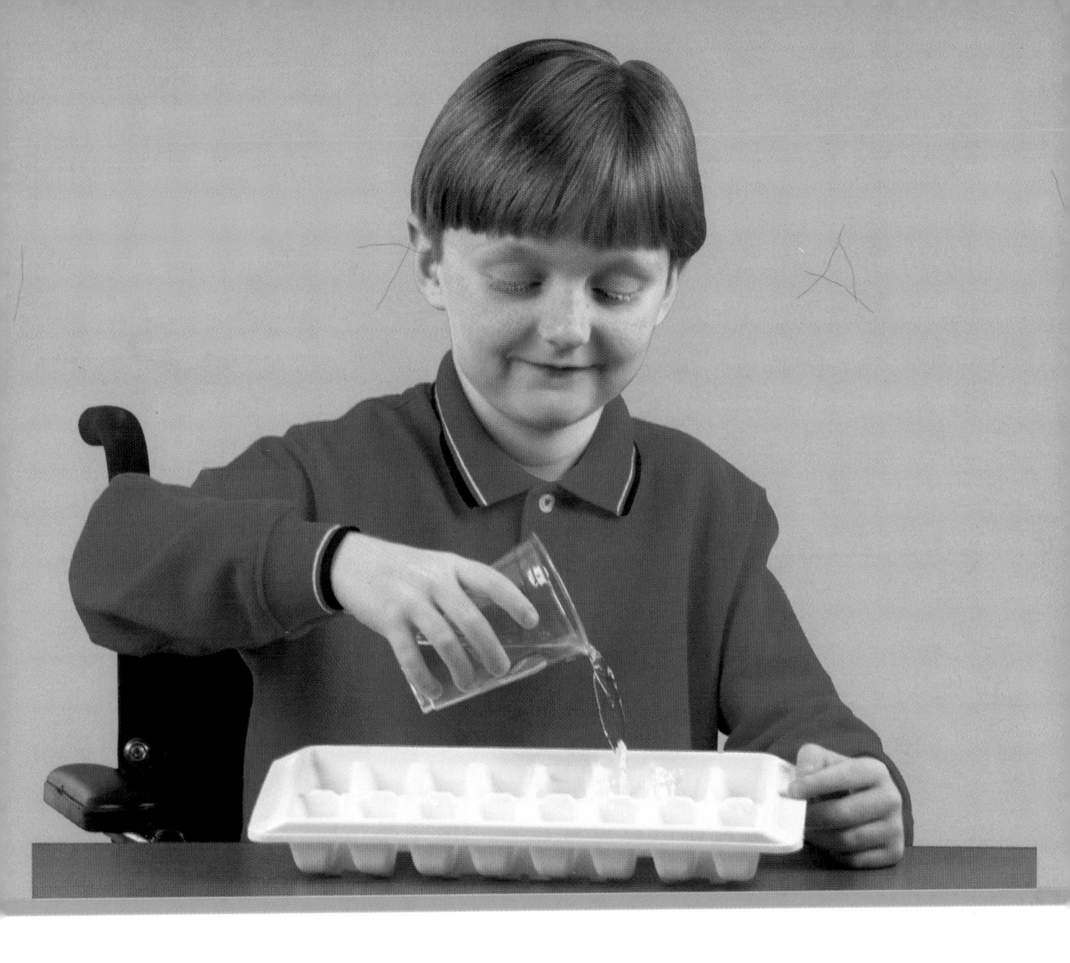

I have water.

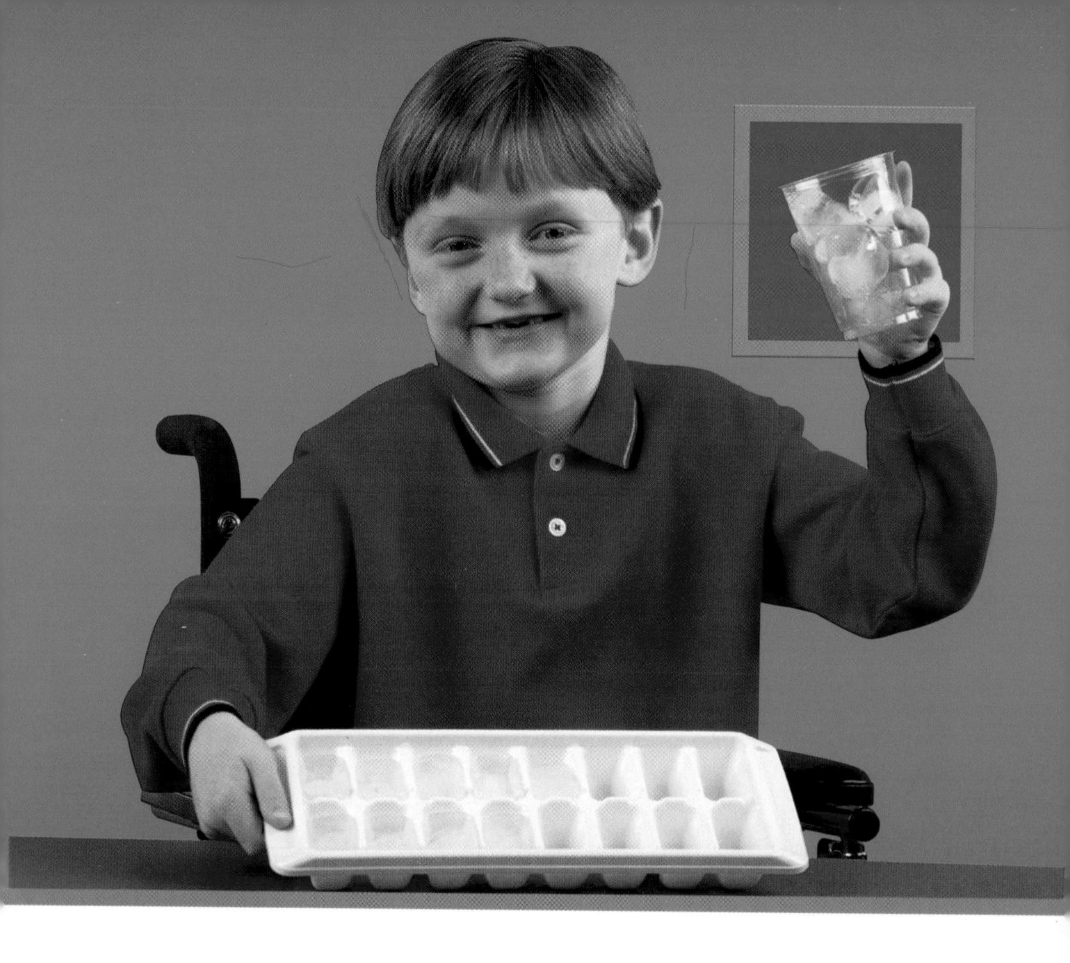

My water is ice!

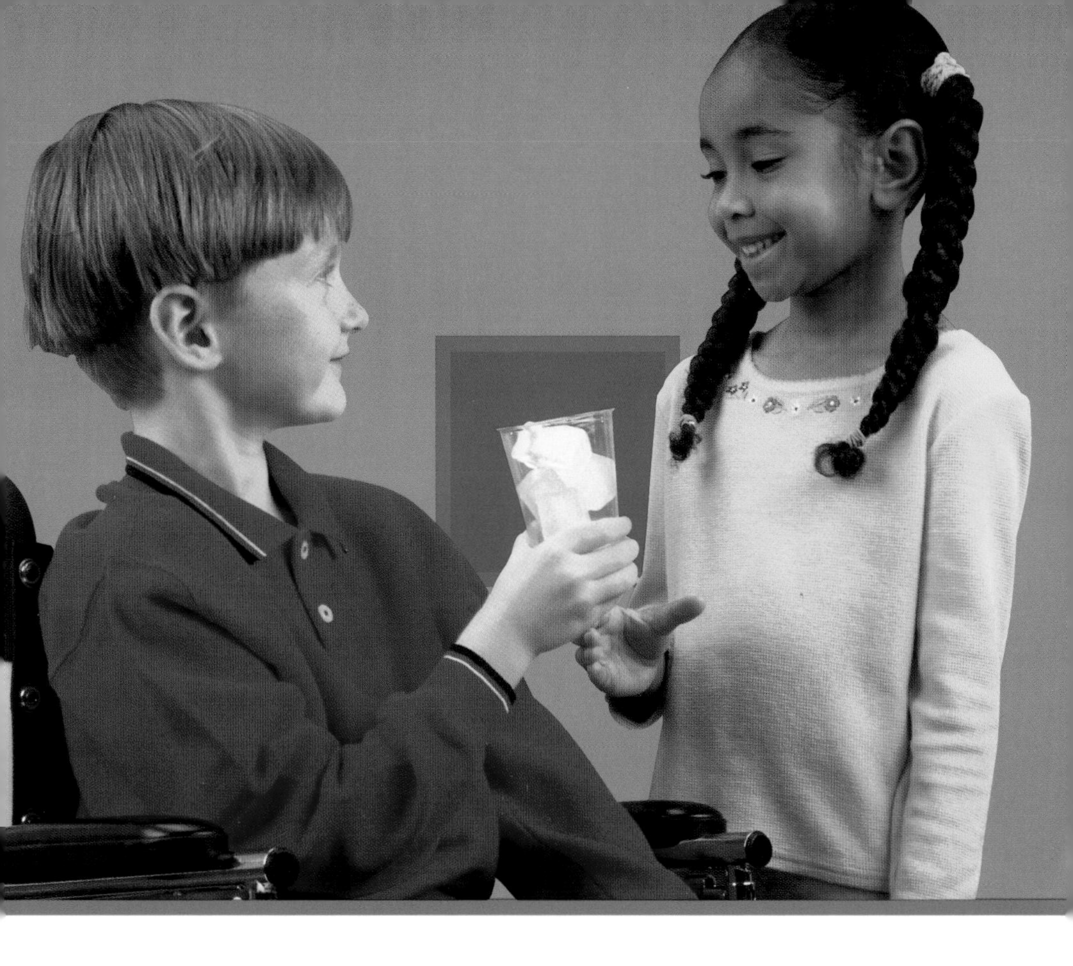

You can have my ice.

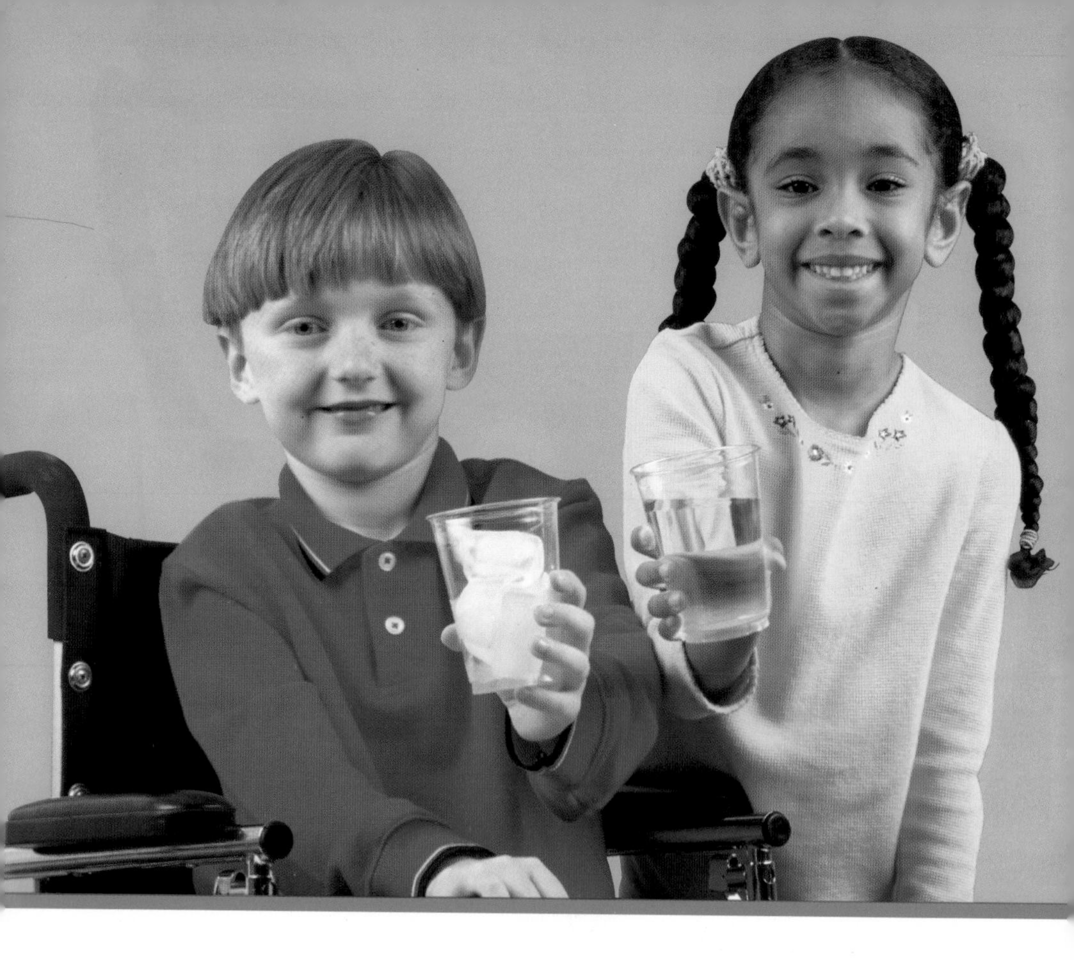

Water!